908 233 519

D0306924

Today's World

The Oil Industry

Malcolm Keir

B.T. BATSFORD LTD
LONDON

CONTENTS

Typeset by Tek-Art Ltd, Kent
and printed in Great Britain by
R. J. Acford Ltd
Chichester, Sussex
for the publishers
B.T. Batsford Ltd
4 Fitzhardinge Street
London W1H 0AH

ISBN 0 7134 5568 3

ACKNOWLEDGEMENTS

The Author and Publishers would like to thank the
following for their kind permission to use copyright
illustrations: *The Aberdeen Journal*, page 34;
Associated Press Ltd, pages 28, 44, 50, 52, 60 (top),
60 (bottom), 66; BBC Hulton Picture Library, pages
16, 56; British Petroleum Company, pages 4, 6, 9, 11,
42; Camera Press, pages 22, 43 (bottom), 49, 64;
E.G.I., page 46; Esso, page 23; Ford, page 18; *The
Guardian*, page 47; Imperial War Museum, page 20;
Oxfam, page 32; Photosource, page 30; *The
Scotsman* publications, page 54; Shell Photographic
Service, pages 3, 5, 8, 10, 13, 14, 15, 17, 40, 43 (top);
The Times, page 58; Topham Picture Library, page
48. Maps and diagrams on the following pages were
drawn by R.F. Brien: pages 22 (Source: *Shell
Educational Service*), 26, 33, 37, 38, 39, 57 (Source:
BP Statistical Review of World Energy 1986), 35
(Source: *The Observer*). Pictures researched by
David Pratt.

THE ORIGINS AND USES OF OIL

Oil is formed from organic substances whose origin lies in the remains of animals and plants which lived in the sea over 50 million years ago. As generation after generation of these tiny creatures died they sank to the ocean floor. Layer upon layer of them was built up and gradually the pressure of the upper layers compressed the lower layers into rock. This rock is called *sedimentary rock*. High temperatures and pressures, and the exclusion of air from the floors of the ocean, started a series of chemical chain reactions going in these sedimentary rocks that led eventually to crude oil (or *petroleum*) being formed. Much of the oil is accumulated in oil "traps" – layers of sponge-like rocks in whose pores the oil lies. Sedimentary rocks are frequently twisted, folded and buckled by the pressures of the earth's crust. When this happens, layers of rock nearer the surface are pushed upwards to form a dome or *anticline*. Oil and gas often accumulate under this dome, trapped by layers of hard impenetrable rock which prevent it moving sideways or upwards. It stays there until its presence is detected and drilling for it starts.

Oil was found in Pennsylvania, USA in 1859, just 70 feet (21.35m) below the ground, but wells today can be up to a mile deep. Prospecting for oil can be a risky business. A certain amount of guesswork is needed about where oil deposits might lie, although geologists and geophysicists have developed many scientific ways of reducing the area of uncertainty today. They study the geological data relating to an area of land via aerial photographs, and prepare maps of a possible oil-bearing region. With the aid of these they can then build up a rough picture of the underlying rock formations and the geological history of the rocks. Certain areas are then chosen for more detailed survey. A drilling area is decided upon and drilling equipment – a rotary drill, drilling bits and hundreds of metres of drilling pipe, together with the supporting derricks, a mechanism known altogether as a drilling rig – are moved into the area. This can take a long time and may involve moving gigantic pieces of equipment over areas of swamp, desert or jungle. Roads may even have to be built. The problems are even worse if floating drilling rigs have to be erected to drill for oil below the sea-bed.

Once drilling starts, the drilling bit cuts down through the layers of rock, and long lengths of

An anticline – the type of land formation under which oil deposits are frequently found. This example comes from Iran.

The semi-submersible oil platform "Sea Explorer" in position in the North Sea. Note the helicopter deck in the foreground.

Life on board an oil rig. A drilling crew prepares to change a drilling bit, Lake Maracaibo, Venezuela.

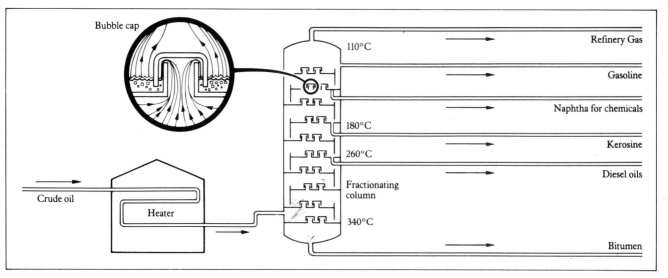

Bubble cap

Crude oil

Heater

Fractionating column

110°C — Refinery Gas

Gasoline

Naphtha for chemicals

180°C — Kerosine

260°C — Diesel oils

340°C

Bitumen

drilling pipe are added to it as it goes deeper and deeper. At the same time, a drilling fluid (called "mud"), a mixture of liquid clay, chemical additives and water, is pumped down through the drilling pipes and then back to the surface, bringing with it the fragments of rock cut away by the bit. The mud is sieved for rock shavings and these rock shavings are then tested for traces of oil. When drilling takes place at sea there are extra problems of keeping the oil rigs stable and upright in the face of high winds and huge

◀ **A diagram in cross-section of a fractioning tower. This shows how, in an oil refinery, the different components (or fractions) of crude oil are separated out by being treated up to different temperatures and then condensed at a temperature just below their boiling point.**

seas. In the North Sea, waves can reach heights of 30 metres or more. Life on board the rigs can be hard, cold and perilous. In March 1980 an accommodation platform (the place where the men lived) serving the oil rigs of the North Sea's Edda field capsized in high seas with the loss of 123 lives.

Once oil is found, the drilling superstructure is removed and an assembly of pipes and valves known as a "Christmas tree" is sealed into the well head (the point from which the oil comes up from under the ground under pressure). This controls the flow of oil through the pipelines to an assembly point where the oil is collected and stored. When it comes out of the ground, oil is known as *crude oil*. Crude oil itself is unusable. It consists of many different liquids, or *fractions*, which have to be separated out and then purified, blended and chemically combined before they can be used commercially (as fuel oil, lubricating oil, etc.). This process is carried out in an *oil refinery*.

Because the fractions which make up crude oil have different boiling points they are separated out by being heated to temperatures high enough to turn them into vapours. Each fraction is then condensed back into a liquid at a temperature just below its boiling point. This takes place in a tall steel tower known as a *fractionating column*. The temperature is kept very hot at the bottom of the column and cool at the top and the inside of the column is divided up into a series of horizontal trays in which the condensed liquids collect. The fractions with the highest boiling points are condensed in the highest trays in the column and are called light fractions. These include refinery gas, gasoline (petrol) and naptha. Those condensing in the lowest trays are the heavy fractions – kerosene (paraffin), gas oil (diesel fuel) and bitumen (tar). The heaviest residues at the bottom are often re-distilled to purify them or to break them up further. There is a much bigger commercial demand for the lightest fractions like petrol, but over half the products of the distillation process are heavy fractions. In order, therefore, to bring the amounts of the different fractions in line with supply, the heavy fractions are chemically changed into lighter ones by being subjected to intense heat. This process is known as *cracking*. A *catalyst* (a substance that helps the chemical change but is itself unchanged in the

A catalytic cracking tower (left of picture) and distillation plant (right) at BP's Dinslaken oil refinery in the Ruhr, West Germany.

process) is sometimes used to assist the process, which is then called catalytic cracking ("cat-cracking").

* * *

Oil has literally thousands of uses today. It is perhaps most important as a fuel for transportation. The 300 million cars in the world drive on gasoline (or petrol) while diesel oil is used for ships and railway locomotives and kerosene for jet engines. Oil is also used directly as a heating fuel in shops, offices, schools and factories. Or it can be used in power stations to drive the turbo-generators from which electricity comes. It is also a lubricant. Lubrication oils minimize the wear and tear on moving mechanical parts, whether these are the delicate parts of a watch or the gigantic pinions of the rollers used in a steel-rolling mill. Car grease reduces engine wear. Oil can also be made into a variety of waxes, like paraffin wax from which candles are made. Paraffin wax is also used for waxed containers like milk cartons. If you have been on holiday recently you will probably recognize the kerosene bottles from camping or caravan sites or on boats and in holiday homes. Bitumen, a heavy oil fraction, is used in road surfacing and for waterproofing and roofing felt.

Another product found along with or near to oil is natural gas. Since 1973, all domestic kitchens in Britain have been supplied with natural gas from the North Sea. This replaced the manufactured gas (or town gas) which at one time you could see stored in large gas holders on the outskirts of many towns. There are still some of them around today. Natural

◄ Concorde, the world's fastest passenger plane, being refuelled at Heathrow Airport. One of the many uses of oil is for kerosene, the aviation fuel.

Detergents are one of the many products that can be made from an oil base. A selection of the huge range of detergents on the market is shown here at a showroom in Liège, Belgium.

Products derived from oil have many uses. They include glues and adhesives (above) and cosmetics (right).

gas, however, is distributed throughout the country by a central pipeline network.

Since the Second World War it has been possible to make entirely new substances both for industry and the home by combining together chemically the various fractions of the distillation process. These substances are called *petrochemicals* and there are literally thousands of them – solvents, industrial alcohols, benzene, nitrogenous fertilizers, detergents, dyes, insecticides, plastics, synthetic fibres, pharmaceuticals, cosmetics and anti-freeze, to name but a few. Oil has thus become an all-important product which enters into almost every aspect of our life today.

THE HISTORY OF OIL

Oil has a very long history. The Greek historian Herodotus tells us that the ancient Babylonians used bitumen from the town of Hit on the river Euphrates as mortar for the walls of their capital. Noah was said to have caulked his ark with tar, and the Byzantines made use of "Greek fire" – flaming arrows soaked in nitre, sulphur and oil, which they hurled at the ships of their enemies. While exploring the Spanish Main, Sir Walter Raleigh came across the pitch lakes of Trinidad in 1595. A century earlier Columbus had found the North American Indians using oil mixed with vegetable dye as a war paint.

For a long time scientists didn't know how to produce oil in large enough quantities to make exploring for it worthwhile; for centuries people had either dug pits for it where it seeped through the earth's surface, or siphoned it off the surface of lakes and streams. In the absence of anything better, people used whale oil for domestic heating and lighting, but by the early nineteenth century the supply even of this was being threatened by over-whaling off the coasts of New England. It was not until the idea of drilling for oil was thought of that it was obtained in really large quantities. The first successful oil strike was made by Edwin L. Drake at Titusville, Pennsylvania on 27 August 1859, and on that day the world's oil industry was born. Oil was to become the most important natural resource of modern times, the key to all the huge technical developments of the twentieth century, both good and bad – the motor car, the aeroplane, the tank and the space shuttle. In 1973 the Shah of Iran referred to it as "the noble fuel", a luxury product too important to be used for anything but the most important purposes, and something that he believed should fetch a luxury price.

From the outset, the oil industry was beset by unpredictability – violent fluctuations in prices and outputs, huge profits and spectacular bankruptcies. Drake's discovery set off the first oil rush. Spectators flooded into Pennsylvania, much as they had done ten years earlier in California when gold was struck. Oil indeed was "liquid gold". Boom towns with exotic-sounding names like Petrolia and Babylon sprang up all along Oil Creek, the site of Drake's

Edwin Drake (right) standing beside his oil well at Titusville, Pennsylvania, in 1859. Drake's was the first successful attempt to drill for oil and marks the beginning of the modern world oil industry.

A study in contrasts. Above, an early American oilfield (Signal Hill, California, in the 1920s) where many oil companies competed to exploit the resources of the oilfield; below, a modern oilfield in the Middle East (Natih, Oman) where one company was given a monopoly over all the oil-development rights.

find, complete with bars, gambling parlours and brothels. Shares in oil claims changed hands across the poker tables in smoke-filled saloons. Even the waitresses, one newspaper reported, were spending part of their earnings buying shares! The early wells were ramshackle affairs and prospecting for oil was a risky business. Many a small-time speculator lost his money. The days of the little man were soon over, however, and by the 1870s the American oil industry was dominated by large capitalist enterprises. The biggest of these was the Standard Oil Company of Cleveland, Ohio, founded in 1870 by John D.

Rockefeller. Rockefeller was a God-fearing, Bible-reading Baptist with a genius for organization and also a determination to impose order on the chaotic oil industry. He quickly established a monopoly over the refining and transportation of oil in the north-east

John D. Rockefeller (1839-1937), founder and president of the Standard Oil Co. of Ohio, USA. Rockefeller was **the most important figure in the early history of the modern oil industry.**

of the United States by ruthlessly eliminating his rivals. His monopoly was not to last for long however. In 1901 the biggest single find in the whole history of the oil industry was made at Spindletop in Texas. So great was the pressure of the oil just beneath the surface that a gusher of oil erupted 60 metres into the air! New oil companies, Texaco and Gulf Oil, were established on the basis of the Spindletop find, and these challenged Rockefeller's monopoly. Rockefeller expanded his refining interests, however, into 30 states, setting up a subsidiary of Standard Oil in each and running them all through his newly formed Standard Oil Trust. His power was only broken in 1911 when the American courts forced him to sell off these subsidiaries. By then he was the richest man in America and also, some said, the most hated. His subsidiaries in due course became major oil companies in their own right and most of them still exist today.

For 20 years America dominated the oil markets of the world; its only serious challenger was Russia. The Russian oil industry had been born in 1873 when for the first time foreigners were allowed to participate in the development of the rich oilfields of the Baku region in the Caucasus. From the outset the Russian oil industry was dominated by the Swedish Nobel family, several of whom were already famous in other fields. Alfred Nobel invented dynamite, and founded the Nobel Prizes for Peace, Literature and Scientific Achievement; his father Emmanuel was the inventor of the torpedo. Together with the Rothschilds, the French banking family who also had shares in the Baku oilfields, the Nobels challenged the Americans in the markets of Europe.

An early filling station (Britain 1922). By the 1920s car ownership was growing in Europe and North America, though mass car ownership was not to occur in Europe until the 1950s.

Oil had also been discovered in Burma and the Dutch East Indies, but there was no way in which the oil companies there could challenge the powerful position worldwide of Rockefeller's Standard Oil until the Royal Dutch Oil Company went into partnership in 1907 with a British company, the Shell Transport Company. Shell had been set up in 1878 by Marcus Samuel, a Jewish bric-à-brac merchant in the East End of London whose business was shipping oriental carvings and sea shells to Europe (hence the name "Shell"). After the merger with Royal Dutch, Shell's shipping fleet was adapted to carry oil through the Suez Canal. In those days oil was a dangerous cargo to carry; it could easily catch fire. Special fireproof tankers had to be built or adapted from other ships to meet the stringent fire regulations of the Suez Canal Company. Royal Dutch, however, could now sell its oil in Europe. It also bought its way

into the Russian oil industry and began to transport Russian oil to the Far East where it could undersell Rockefeller.

By 1900 oil had also been discovered in Venezuela and Mexico and, not for the first time, the oil companies faced the prospect of a glut and a collapse of prices. At the last minute they were saved by perhaps the most important thing to happen in the whole history of the oil industry, the invention of the

motor car. The first motor car was manufactured commercially in Germany in 1896 by the Daimler-Benz Company but it was the development of the Model T car, the first mass-produced car in history, by Henry Ford in Detroit in 1908 which transformed the fortunes of the oil industry. By 1912 there were over a million cars in the USA. The demand for oil now switched from kerosene (a heating and lighting agent, itself now increasingly displaced by electricity) to petrol. At this time, too, ships were also turning over to fuel oil.

As the threat of war loomed before 1914 the advantages of oil-powered warships were not lost on the Great Powers, particularly Britain, the largest naval power of the day. Ships powered by oil were faster than coal-fired ships since they dispensed with heavy coal bunkers and also had no need to return to port for re-coaling. In the uncertain atmosphere preceding the war the British government was anxious to find a supply of oil entirely under British control. Luckily, oil had been discovered in southern Persia in 1908 by a group of British prospectors, and in 1913, at the prompting of Winston Churchill, the First Lord of the Admiralty, the British government bought a controlling interest in the new Anglo-Persian Oil Company. Even though only a limited amount of Persian oil was used in the war, it was an astute move, which gave the British government an interest in the Middle East for many years to come.

Oil proved crucial in the First World War. This was the first war ever to be fought by tanks and aeroplanes, and Germany's relative lack of oil proved a crippling handicap. The British, the Americans and the French had 150,000 trucks and Allied soldiers could be moved quickly up to the battle lines. (British troops actually arrived at the battle of the Somme in 1916 in London taxis!) The Germans meanwhile relied throughout the war on horse-drawn transport and many horses were killed. Lord Curzon, the British Foreign Secretary summed up the situation well when he said that the Allies "floated to victory on a wave of oil".

So vital was oil now considered to be that there were widespread fears after the war that there would not be enough to go around, especially as Russian oil exports to western Europe were cut off after the

The production line at Ford's Dagenham plant in the 1960s.

Russian Revolution in 1917. Anxious to avoid being too dependent on American oil, the two major west European powers, Britain and France, now started looking to the Middle East as a possible new source of oil. Under the San Remo Agreement of 1920 they had been given control over Iraq and Syria, two of the successor-states which arose out of the ruins of the Turkish Empire. Oil had been discovered in Iraq before the First World War, and the British and French governments now made sure that their own national oil companies, Shell, Anglo-Persian and Petroles Commerciale de France, took over the former German shares in the Iraq Petroleum Company. This had been established in 1912 to develop Iraq's oil reserves.

The Americans, not to be outdone and themselves

worried that their own oil might run short, also demanded a share in the new oil wealth of the Middle East, claiming that the principle of "the open door" (i.e. equal access to all) should apply to the new oil finds. American claims could not easily be ignored (the Americans after all had played a decisive role in the First World War) and in 1928 five American oil companies were allowed to buy their way into the

Iraq Petroleum Company. The expected oil shortage didn't materialize, however; instead, huge new discoveries were made in eastern Texas and Venezuela. Once more the threat of another oil glut loomed on the horizon. The problem for the oil companies was again how to prevent a collapse of prices and profits. Their solution this time was to organize a producers' cartel to limit production and keep up prices. In 1928 the chairmen of the three leading oil companies – Walter Teagle of Standard Oil (later Esso), Sir Henri Deterding of Shell and Lord Cadman of Anglo-Persian – met together at Achnacarry Castle, the shooting lodge of a Scottish aristocrat, Lord Cameron, and made an agreement to share out the world market between them. Production was to be limited by these oil companies, and they agreed to supply each other with oil at fixed prices whenever any of them ran into shortage. Four other companies – Mobil, Texaco, Chevron and Gulf Oil – were brought into the agreement in 1934.

With the onset of the Second World War the fear of a glut once more came to an end. Again it was Germany, and now also Japan, who faced a crucial shortage of oil. The Allies controlled the oilfields of the Middle East and also the shipping lanes of the north Atlantic, although German submarines were to menace the tanker traffic from the USA to Britain until the end of 1943. The Germans were forced to manufacture oil synthetically from coal, and one of their major aims when they attacked Russia in 1941 was to capture the Caucasian oilfields. The *blitzkrieg* – the series of lightning attacks waged on selected points by the German armies against Poland in 1939 and France in 1940 – was designed to get the fighting over quickly so that scarce German oil reserves could be conserved. In the Far East, the Japanese launched their first attacks of the war (in 1941) against Malaya and the East Indies in order to try to seize the crucially important rubber estates and the oilfields. In 1944, without huge and guaranteed supplies of oil the Anglo-American invasion of western Europe could not have taken place. In order to supply the Allied landing forces, a huge oil pipeline was laid under the English Channel from Britain to the Normandy

Troops of the 2nd Battalion, Royal Warwickshire Regiment being transported to the front by buses of the London General Omnibus Co. (First battle of Ypres, 6 November 1914.)

**The Anglo-American invasion of Europe ("D-Day").
Army lorries and equipment being put ashore on the
Normandy landing beaches, 8 June 1944. Once ashore,
the invasion forces could move more swiftly with the
help of motorized troop carriers. But these had to be
kept supplied with petrol.**

beaches. This was called PLUTO, which stood for
"Pipeline Under the Ocean".

The growing importance of oil in the world
worked to the advantage of the large oil companies.
The period after the Second World War saw them at
the height of their powers. Instead of the expected
world economic slump (as had occurred after the
First World War) the major world economies went
through an unparalleled period of expansion and
prosperity. Private car ownership boomed and the
demand for consumer goods increased as never
before. A whole range of new products, known as
petrochemicals, had also been developed during the
war to meet war-time shortages. Detergents,
fertilizers and nylon were just a few of these. These
were all by-products of oil, and the oil industry was
thus set to enjoy a new and uncharted period of
expansion into a wholly new field. True to its history,
however, the post-war years were to witness as many
convulsions and turnabouts of fortune in the industry
as any of the previous 90 years had seen.

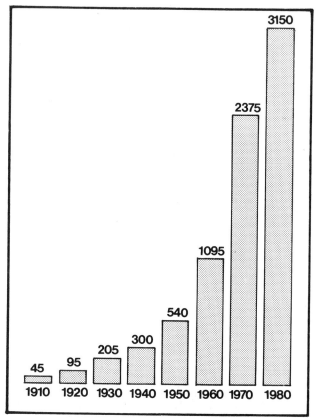

**World crude oil production 1910-80 (in millions of
tonnes).**

THE ECONOMICS OF OIL

Oil today is the biggest business in the world. The financial resources and the level of activity involved in exploring, refining and selling oil vastly exceed that of any other industry. Of the ten largest companies in the world, six are oil companies. The turnover of the largest of them, Exxon (the former Standard Oil Company of New Jersey) is greater than the gross national product of Switzerland or the annual defence budget of NATO.

For 25 years after the Second World War a number of powerful oil companies dominated the world oil industry. Known as "the Seven Sisters" (Exxon, Mobil, Gulf, Texaco, Chevron, Shell and BP) they controlled all the stages of production and distribution, from exploration through to the refining, transportation and marketing of oil. Ever since the days of John D. Rockefeller the oil companies as a whole had had a reputation for ruthless, unscrupulous profit-seeking, and rigging the market at the expense of the consumer. Their huge assets, the secrecy that surrounded their operations and their seeming ability to switch their funds and operations from one part of the world to another at will made them objects of suspicion and resentment among both producers and consumers. Since most of them were American, and since the USA had always thought of itself as the world's leader in oil technology and "know-how", their activities also fuelled anti-American sentiment.

"Put a Tiger in your Tank" – Esso's advertising slogan in the 1960s.

The extent of their control was impressive. In 1972 the Seven Sisters owned 62 per cent of all the world's oil assets and were responsible for 57 per cent of all new oil investment. Most importantly, they also controlled 90 per cent of the Middle East's exported oil. Although in theory they were competitors, in practice they cooperated very closely with each other, one company "swapping" products in which it had a surplus (say kerosene) for another in which it was short (lubricating oil perhaps). This cooperation was especially important given that the demand for different types of oil varied considerably across the globe. Forty-four per cent of all the oil needed in the USA was for petrol, whereas over 50 per cent of the Japanese demand was for diesel oil and lubricants.

The oil companies also tried to extend the range of their operations from refining right through to marketing. Those, like Mobil or Shell, who were strong at the marketing (or "downstream") end of the business tried to buy their way into more "upstream" activities like oil refining, while those who were well established at the "upstream" end, like BP, Texaco and Standard Oil, tried to acquire marketing outlets "downstream". There were also a number of arrangements under which several companies agreed to set up subsidiaries to develop new oilfields. The best known of these was Aramco (the Arab-American Oil Company) which was established in the 1950s by Exxon, Socal, Texaco and Mobil to develop new oil finds in Saudi Arabia. The oil companies also acted as middlemen between producers and consumers by holding oil stocks for when they were needed. In particular, they helped to

THE SEVEN SISTERS

Exxon (or **Esso** as it has been called in Europe since 1926) is the direct descendant of Standard Oil of New Jersey. Today, it is the largest oil company in the world, with interests in all the world's major producing regions. It is also involved in every kind of energy resource from oil and natural gas through to coal and even uranium. Exxon hit the headlines in the 1960s with its advertising slogan "Put a Tiger in your Tank!"

Shell is the second largest of the world's oil companies. An Anglo-Dutch company, it was formed in 1907 as the result of a merger between a British shipping company, Shell Transport and Trading, and the Royal Dutch Oil Company. Its board of directors today is split 4:3 in favour of the Dutch but its main headquarters is in Britain, on London's South Bank. Until after the Second World War it was strongest at the marketing end of the oil business, but developed extensive new oilfields of its own in Venezuela in the 1940s and Nigeria in the 1960s.

British Petroleum is a wholly British-owned company. Founded in 1908 as the Anglo-Persian Oil Company, it became the Anglo-Iranian Oil Company in 1935 and BP in 1954. It has been long involved in the Middle East, first in Persia, then in Iraq and finally in Kuwait. The British government had a controlling interest in it from 1913 until 1979. It has often been embroiled in Middle East politics, as in 1951 when its refinery at Abadan on the Persian Gulf was seized by the Iranian government.

Socal (Chevron) The name Socal stands for Standard Oil of California. Founded as the Pacific Coast Oil Company in 1879, the company was bought up by John D. Rockefeller in 1900 after a savage price war with Rockefeller's own Standard Oil Company. It was one of the companies which developed the huge oil reserves of Saudi Arabia in the 1930s, along with Texaco, with whom it formed a joint production and refining company known as Caltex. It trades today under the brand name of Chevron.

Texaco is one of the two big oil companies that grew up on the basis of the huge Spindletop find in Texas in 1901. Known originally as the Texas Fuel Company, it first sold oil to the sugar planters of the American Deep South who used it to heat their sugar boilers. It expanded in the inter-war years by buying up other oil companies, and took a leading part in developing the Saudi Arabian oil fields. Today it also has interests in Trinidad and Venezuela.

Gulf Oil is the second of the Spindletop oil companies. Today, however, its headquarters are in Pittsburgh, Pennsylvania. It went into partnership in 1934 with the Anglo-Persian Oil Company to develop the Kuwait oil fields, and became a major force in the Middle East oil business. Stronger in production than in refining or marketing, it sold much of its Middle Eastern oil to Japan and the Far East. In 1984 it merged with Socal in what was the biggest merger in American corporate history.

Mobil was formed in 1931 as the result of a merger between two former Rockefeller companies, Socony (Standard Oil of New York) and the Vacuum Oil Company. It later became Socony Mobil, and took the name Mobil in 1966. It has a strong position today in the Far Eastern oil market.

even out seasonal fluctuations in the demand for oil. The demand for petrol always goes up in the summer months, while the demand for heating oil rises in the winter.

The oil companies have always claimed that their activities help to smooth out the fluctuations in the oil market, and therefore that they bring benefits to both producers and consumers. To some extent this is true, but it was their monopolistic control over oil prices which traditionally aroused the greatest hostility. The story of this goes back to the 1920s. Under the Achnacarry Agreement of 1928 the companies agreed among themselves to sell their oil at a fixed price (known as the "posted price"). This was related to the cost of producing American oil in the Gulf of Mexico. Added to this price, however, was a surcharge based on the theoretical cost of moving oil from Galveston, Texas to wherever its destination was in the world. This was payable regardless of whether the oil had actually come from the USA; it might simply have been shipped from the Persian Gulf to India. This system of pricing, known as the "Gulf Plus System", was designed to protect the exports of American oil against competition from the cheaper oil of the Middle East. Since all the oil producers, including those in the Middle East, were paid a royalty on their oil based on the posted price of American oil they were happy with the system. It was the oil consumers who suffered.

Opposition to the entrenched position of the big oil companies in the 1950s came from three sources. First, from a number of smaller American oil companies, sometimes called the "independents", who tried to break into the oil markets of the big companies by undercutting their prices. The big companies retaliated by launching massive advertising campaigns. They also tried to cut their costs by sharing things like pipelines and exploiting to the full the advantages of using large "supertankers" to transport oil round the world.

Another challenge came from Europe. The stranglehold of the big Anglo-American oil companies (five of the "Seven Sisters" were American-owned and the others were British and Anglo-Dutch) alarmed a number of European consuming countries, particularly the French and the Italians. They therefore set out to find their own oil supplies. In 1953 the Italian state oil company ENI (Ente Nazionale Idrocarburi) was established; this was followed in 1966 by the French company Elf-ERAP. The Italians, under their energetic chairman Enrico Mattei, tried to outflank the big oil companies by obtaining oil from the Libyans, the Iranians and also the Russians. But the state oil companies could only hope to buy their way into *new* oil areas. The existing oilfields, especially the rich areas of the Middle East, remained firmly controlled by the big oil companies.

It was, in the end, a quite different challenge, that from the producers themselves, which was to break the power of the oil companies. There was no love lost between the oil companies and the oil producers. For over 50 years the economies of the oil-producing countries, particularly those in the Middle East and the Caribbean, had been controlled by foreign oil companies over whom the host governments had little control. These companies could decide how much oil they wanted to take out of the ground and at what price it was sold. They could also do this without consulting the producing countries themselves. The oil companies had long ago, as far back as the 1920s in some cases, negotiated agreements with the governments of the producing countries that gave them legal control of the oil for as much as 100 years ahead. They had this power because the oil producers themselves lacked any skilled manpower or oil equipment of their own with which to exploit their own oil resources.

To make matters worse, the oil that these foreign companies controlled was the one thing which many producing countries depended on for all their foreign earnings. It was a humiliating position in which the producing countries found themselves, and not surprisingly they resented this deeply. Their position was made worse still by the fact that for 30 years after the Second World War there was more oil being produced than could be consumed, even at the high levels of consumption prevailing after 1945. This was because huge new discoveries of oil were made in Saudi Arabia and the Gulf states in the 1950s. The producing countries therefore found themselves competing against each other to sell their oil, while the oil consumers (chiefly the advanced industrial countries like Britain) for a long time got the benefit of low oil prices.

There were also differences of interest between the oil producers. Desert states like Saudi Arabia and the sheikhdoms of the Persian Gulf were rich in oil but

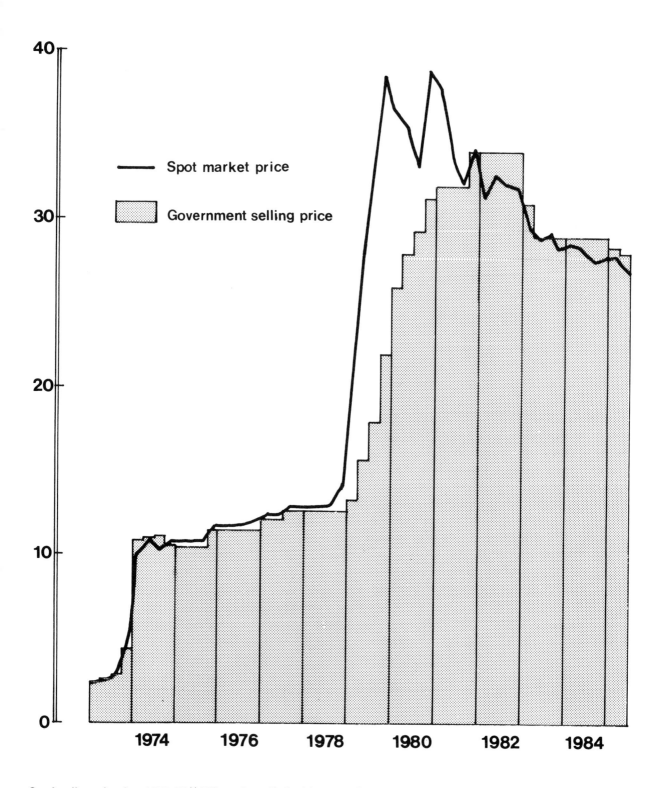

Crude oil production 1974-85 ($ US per barrel). Arabian Light Crude is the best quality Saudi Arabian oil. The *Government selling price* was the price recommended by the Saudi Arabian government. The *spot market price* was the price of small consignments of oil on the Rotterdam oil market, often a good indicator of the short-term (even day-to-day) demand for oil.

had small populations. These countries had little need for imports on any scale and therefore they had little incentive to extract all the oil that lay under the ground. Some of them, like Kuwait, had large but not unlimited supplies of oil and were anxious not to sell it all at once. Another group of oil producers, however, was in a totally different position. Countries like Nigeria, Iran and Algeria had large and often poverty-stricken populations. They saw in their oil wealth a way of rapidly escaping from their poverty. These countries all needed to import goods from the industrialized world on a large scale to develop their economies. Their aim was to raise the living standards of their impoverished populations within a generation or two. Their interests therefore lay in expanding their oil production as fast as possible and, out of the profits from oil, building an economy that was not just dependent on oil alone but diversified into other industrial areas as well.

Among the oil producers, one country, however, stood out – Saudi Arabia. So great were its oil reserves that alone, by raising or lowering its output, it could virtually decide what the price of oil would be, whatever all the other producers thought or did. As an oil producer, Saudi Arabia had an interest by 1970 in seeing the price of oil rise above the relatively low levels of the 1950s and 1960s, but not to such a height that it drove its own customers away, since there was little else that it could rely on apart from oil.

Whatever their individual level of output however, all the oil producers were united in wanting a higher return per barrel on the oil they produced. Venezuela led the way in pressing for this. In 1948 she forced the oil companies to share their profits on a 50/50 basis with her. In 1950 the Aramco consortium was required to pay a 50 per cent tax to the Saudi government on its oil earnings (though it was able to claim this back in tax relief in the USA). But advances like these still fell short of giving the producers what they really wanted, which was control over the actual *level* of production and prices. Just how difficult it was for the producers to take control over their own oil was shown in 1951 when the Iranian government suddenly and unexpectedly nationalized the assets of the British-owned Anglo-Iranian Oil Company and seized its huge refinery at Abadan on the Persian Gulf. Anglo-Iranian, however, controlled all the sales outlets for Iranian oil. It withdrew its technicians from Abadan, thus bringing the output of

the refinery to a standstill. At the same time it persuaded its fellow oil companies to boycott Iranian oil. The Iranian government thus found itself in a difficult position – it now had control over its own oil but there was no one to whom it could sell it!

As oil surpluses rose in the late 1950s the big oil companies, without consulting the producers, suddenly decided to cut the posted price of oil. Under the 50/50 agreements, which had now become widespread, and also the arrangement under which oil company taxes had been linked to the posted price of oil, the oil producers now found that their oil incomes had been cut and with it their ability to buy foreign goods. Five of them – Venezuela, Iraq, Iran, Saudi Arabia and Kuwait – therefore got together in 1960 to resist any further price cutting without their consent (and indeed to restore the price cuts of 1959-60). They formed an organization called OPEC – the Organization of Petroleum Exporting Countries. By 1973 OPEC had 13 members. In addition to the founder members, these were: Qatar, Indonesia, Libya, Abu Dhabi, Algeria, Nigeria, Gabon and Ecuador.

At first, little notice was taken of OPEC by the oil companies. By 1970, however, a crucial change had taken place in the oil market; for the first time since the Second World War the oil surplus was replaced by an oil shortage, thanks largely to the continuously rising demand of the industrialized West. In particular, the USA had shifted from being an oil exporter into becoming an oil importer and the West as a whole had come to depend almost entirely on imports of Middle Eastern oil. Market conditions now favoured the oil producers, but it still required some country to strike the first blow against the oil companies. That country was Libya. In 1969 a military coup had brought to power an extreme nationalist, Colonel Muammar Qadhafi. Qadhafi was bitterly opposed to the Western oil companies and he now inflicted a decisive and irrevocable defeat on them.

Although only a small country in terms of population, Libya was in a uniquely strong position in relation to the oil companies. Since the Arab-Israeli war of 1967 and the closure of the Suez Canal, the Western oil-consuming countries had become increasingly dependent on Libyan oil. Libya had a small population and therefore had no great need to sell all her oil immediately. Qadhafi also took advantage of the presence in Libya of a number of small independent oil companies who, unlike the big companies, were entirely dependent on Libyan oil. In May 1970 Qadhafi threatened to cut off their supplies unless they agreed to cut their output immediately, to raise their posted price (and thus the amount of revenue they paid to Libya) and to alter the 50/50 profit share in Libya's favour. The major oil companies, taken by surprise and divided among themselves about what action to take against Libya, were eventually forced to agree to the same terms as the independents. Other members of OPEC quickly followed Libya's lead. In the early 1970s many producer countries went even further than this and completely took over production from the oil companies, leaving them with the job of simply selling the oil. Iraq nationalized its oil industry in October 1972 and was followed by Iran in May 1973. By 1980 Kuwait, Bahrain, Qatar, Venezuela and Saudi Arabia had also obtained complete control over

their oil industries.

Things were not left here, however. Prior to 1970 the revenues of the oil producers had been linked to the posted price of oil as set by the oil companies. OPEC now started to dictate the posted price, and raised it sharply. In October 1973 the price was pushed up by 70 per cent and this was followed by a further 130 per cent increase in December. In less than a year therefore the OPEC countries had managed to seize control of prices from the oil

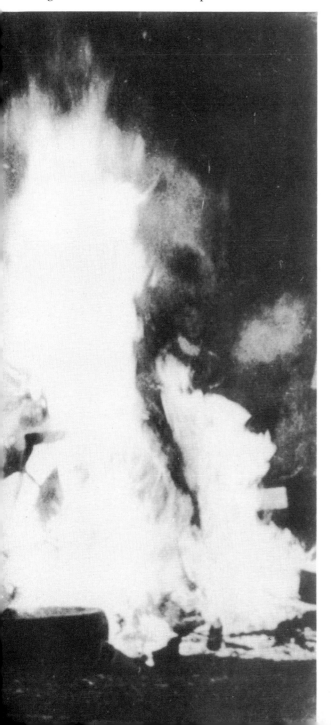

companies in one of the most dramatic shifts of power the industry had ever seen. At the same time, OPEC started to impose production cuts. These cuts were if anything even more damaging to the oil consumers than the price increases. In the days when the companies controlled output they always made sure that the consumers got as much oil as they needed. Now, however, the consumers were made to go short of oil if necessary. The view of the OPEC countries was that the supply and price of oil had to be arranged to suit *their* needs rather than the convenience of the oil consumers. This was a new and profound change of attitude. For the oil consumers things looked grim. For them it was, said Sir David Barron, the chairman of Shell, "like looking down the barrel of a gun". The consumers attempted to retaliate but it was a weak and ineffective attempt. The USA tried to organize a "consumers' OPEC", the International Energy Agency, which would have brought the consumers together in a united front against OPEC in order to force the price down. But many consumer countries preferred to find alternative sources of oil outside the Middle East, and others like France attempted to make their own independent deals with individual Arab countries.

By the end of 1973 oil prices had trebled. They rose again in 1979 when OPEC imposed a further 10 per cent price increase and then, dramatically, when the oil market was deprived of Iranian oil following a *coup d'état* in Iran in which the pro-Western Shah was deposed. The war which followed between Iran and Iraq also led to the loss of Iraqi oil exports. At the end of 1980 the price of crude oil on the world market was $38 per barrel compared to the $2 per barrel it had been in 1974! Even if all other prices had remained at their 1974 levels instead of doubling in six years (the car which cost £2500 in 1974, for instance, cost over £5000 in 1980) oil prices would still have risen to *ten times* what they were in 1974!

* * *

What were the effects on the rest of the world of these huge price increases? For the oil-producing countries, particularly those in the Middle East, who were the largest producers, the prospect of untold

Some Americans reacted strongly to the shortage of petrol ("gas") in the USA in the late 1970s. In Levittown, Pennsylvania, in June 1979, cars were set on fire as demonstrators rioted against petrol shortages.

wealth seemed to lie on the horizon. With it came also the prospect of a massive redistribution of the wealth of the world in favour of the poor countries (or some of them) and against the rich countries who had for so long, in the view of the poor countries, exploited the Third World.

The desert kingdoms of the Arabian peninsula and the Persian Gulf with their small populations had no great need for imports. They therefore invested much of their oil wealth abroad. They also diversified their economies into oil-related industries like petro-chemicals so they would have an industrial base if ever their oil ran out, or if the price of oil collapsed. The governments of the Gulf states could also afford to build brand new luxury hotels, airports, motorways and office blocks in their own countries, and many formerly poverty-stricken Arab towns now became transformed as if by magic into ultra-modern twentieth-century cities. The lives of ordinary people, too, often improved greatly in these states as modern schools and hospitals were built. In Kuwait the illiteracy rate dropped 30 per cent, and for the first time women were able to have their babies in well-equipped hospitals. Women benefited in other ways too. Even though they were still segregated from men at work, they were able to qualify as doctors and teachers and for the first time to take up work outside the home.

Other oil-producing countries like Nigeria or Iran had less freedom than this, however, in the way they spent their oil revenue. With large and growing

A street scene, Kuwait City. Note the modern buildings which the new oil wealth of the Middle East made possible.

populations to support, they were obliged to spend much of their oil wealth on essential imports of food and manufactured goods. But they too were beneficiaries of the oil boom.

In the consuming countries, however, things were very different. The advanced industrial countries of western Europe and North America who had little oil of their own or who, like the USA, had some oil but were not completely self-sufficient, were now forced to pay up to ten times more for their oil over a period of just seven years. Since oil enters into the price of almost everything that is manufactured, "oil inflation" became a major factor in the economic life of the industrial countries. Rising fuel costs led to bankruptcies, unemployment and an economic slump. Ironically, the higher price of oil found its way back to the Middle East through the sharp rise in

the price of the manufactured imports which the oil producers themselves were now buying from the West. Because they now had to pay more for these imports they found that their oil income was able to buy less than they had expected.

Things were worst of all in the very poorest countries of the world – in Asia, Africa and Latin America – the majority of whom produced no oil of their own. Here one of the main uses of oil was as kerosene for cooking. But now poor families in the rural areas were forced instead to use firewood. This led to much deforestation and to longer and longer journeys each day just to find wood. Village pumps and electricity generators ran out of oil, and deep-sea fishermen in places like the Maldive Islands found their livelihoods threatened when they could no longer afford diesel oil for their fishing boats. These countries stood on the brink of economic disaster.

* * *

Sooner or later, the rise in oil prices was bound to come to an end. Oil is a commodity which is traded on the world market. Like any other – tin, copper, grain, sugar – it will sell at a price at which at any one time the demand matches the supply. If the supply increases relative to the demand, or the demand falls relative to the supply, then the market price falls. By 1980 prices had peaked and by 1985 they were again down to their 1979 levels. There were a number of reasons for this. First, demand fell in the advanced industrial countries as the growth in production was halted and in some instances reversed. Gradually these countries also turned to cheaper alternative sources of fuel – coal for instance. They also learned to use oil more economically. Some countries like France went ahead faster than before with ambitious nuclear-power programmes. But the biggest change was the emergence of new sources of oil supplies.

The high price of oil in the 1970s made it profitable for the oil companies to start drilling for oil in areas which had previously been too costly to develop, like the North Sea and northern Alaska. These new oilfields, which were all outside the Middle East, lay beyond the control of the OPEC countries. Hence, OPEC's control over the supply and price of exported oil began to disintegrate. Oil production in these new areas grew rapidly, particularly in the North Sea; so much, indeed, that Britain had become the world's fourth largest oil exporter by 1985. The north-east of Scotland, especially the area around Aberdeen, and also the Shetland Islands off the north coast of Scotland, where much of the oil from the North Sea was brought ashore by pipeline, enjoyed a "mini boom" as the oil companies moved in and built new refineries and terminals. Local unemployment in what had been a farming and fishing area fell dramatically as new jobs were created in the building industry and retailing. The oil industry created jobs in other parts of Scotland too – on Clydeside, for instance, where many of the oil rigs were built in what had formerly been shipyards.

Gathering wood in Tanzania. Wood still supplies 80 per cent of the fuel needs of poor countries. The rising price of oil in the 1970s caused even heavier use of wood and led to the depletion of much woodland in the developing world.

The OPEC countries now faced a crisis. The only way in which oil prices could be kept up was if they themselves cut their own production. This was easy enough for Saudi Arabia and the Gulf states, but the OPEC countries with large populations needed all the oil revenue they could get. They believed that the only way they could keep up their oil incomes was to sell more oil. But, of course, the more oil they sold, the lower the price fell on the world's markets! For a long time Saudi Arabia, by far the largest oil-exporting member of OPEC, tried to offset this tendency for the price of oil to drop by cutting its own output as others increased theirs. Between 1982 and 1985 Saudi Arabian output fell from 11 million barrels per day to just two million barrels per day – at one point, less than Britain's output. In the end however these cuts put her own economy in jeopardy

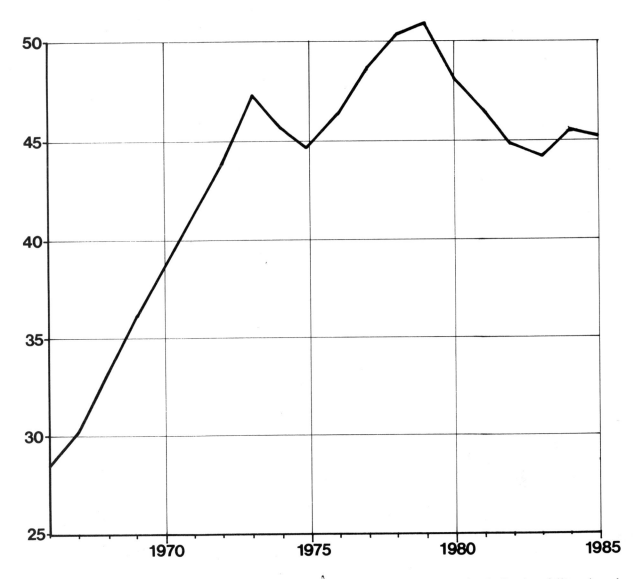

Consumption of oil 1966-85 in the non-communist world (in thousands of barrels per day).

and at the end of 1985 she abandoned her policy of production restraint. As a result, oil prices dropped below $10 per barrel for the first time since 1973. The OPEC era seemed to be over.

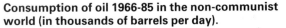

You might think it is a good thing that oil prices have fallen so much. And so it is for the oil-consuming countries, both poor and rich, although there are one or two countries, like Britain, which are in the odd position of being both rich countries and also oil exporters. As a rich country, Britain stands to gain from any revival of trade and prosperity among her rich neighbours as a result of oil prices falling, but she will also lose much of her income from oil exports now that oil prices are lower. The oil boom in Scotland has now come to an end, as the fall in prices is making any further exploration of the North Sea uneconomic for the oil companies. Scotland is perhaps only the latest tragic casualty of the "boom-bust" cycle which has been so much a part of the history of the oil industry.

Scotland is not the worst casualty, however. Far more devastating has been the effect of collapsing oil prices on countries like Mexico and Venezuela which are either relatively poor oil exporters or countries which depend almost entirely on their oil to buy imports. Venezuela's oil sales in 1984 represented 90 per cent of her total export earnings and Mexico's 66 per cent (the comparable figure for Britain is 10 per cent). For countries like these, collapsing oil prices mean seriously declining living standards. Mexico, with her population of over 60 million, is in special difficulty. The situation is serious too for the foreign banks who, perhaps unwisely, lent money to these countries in the 1970s when oil prices were high, for they now stand to lose their loans. Banks, of course, always take a risk when they lend money. They do so in the expectation of making profits. But bank lending has a wider significance since if a number of poor countries were to fail to repay their debts a

The collapse of oil prices in the mid-1980s brought an end to the Scottish oil boom. Here, redundant drilling rigs stand idle in the Cromarty Firth. October 1986.

number of leading European and American banks could collapse. In the 1930s a banking collapse helped to bring about a worldwide economic depression. The same thing could easily happen again.

The future of oil prices today is highly uncertain. If prices continue to fall, high cost production regions like the North Sea will eventually go out of operation altogether, leaving the field free once more for

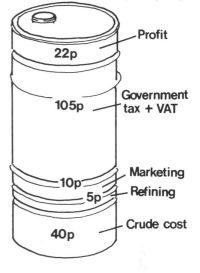

$20 Crude price per barrel
182p Petrol price per gallon

- Profit 22p
- Government tax + VAT 105p
- Marketing 10p
- Refining 5p
- Crude cost 40p

Where your money goes: the price of a gallon of petrol in Britain, January 1986.

OPEC, and perhaps for a repeat of the oil price instability of 1973-86. If, on the other hand, OPEC reintroduces production restraints in the next few years, the fall in prices could be halted and some part of the North Sea (and also the Mexican) oil industry saved. One thing is certain: that in the long run the demand for oil will continue to rise, if only from the rapidly industrializing countries of the Third World with their large and growing populations. Whether prices will also rise depends upon how much new oil can be found to meet that demand, and at what cost.

THE GEOGRAPHY OF OIL

Oil-bearing rocks are found in many parts of the world, in what are known as *sedimentary basins*. They are also found on the continental shelf – the area of shallow water which extends round the coastlines of the major continental land masses (though "shallow" here means anything up to half a mile deep, compared to depths of five or six miles in the deepest parts of the oceans). Indeed, as onshore oilfields have been exhausted, so the search for new oil has switched to offshore locations. In 1980 offshore oil production accounted for more than 20 per cent of total world production.

Most of these potential oil-bearing areas, however, are unlikely ever to be explored, because they lie in remote or difficult terrains – deserts, jungles or mountains. Most of the easily extractable oil is found in a few areas of the world. The world's largest oil producer today is the Soviet Union, responsible for over 20 per cent of the world's output. Relatively little of this is exported, however; most of it is either consumed inside the Soviet Union or sold under rather special terms to Russia's East European neighbours in the COMECON, the Communist economic bloc of countries. The Soviet Union is, however, a major exporter of natural gas, particularly to western Europe. The world's second largest producer is the USA. The oil industry was born in the USA, in Pennsylvania, but by the turn of the century there were other fields in production in California, Texas, Louisiana and Oklahoma, which is where most of America's oil still comes from today. In the 1980s huge new oilfields were discovered in northern Alaska. The USA was once a major exporter of oil; today, however, despite repeated attempts to become so, she is no longer self-sufficient in oil and has to import from the Caribbean and elsewhere. At one time, in the 1970s, she was heavily dependent on Middle Eastern oil.

Areas of actual or potential oil and gas production, on-shore and on the continental shelves and slopes.

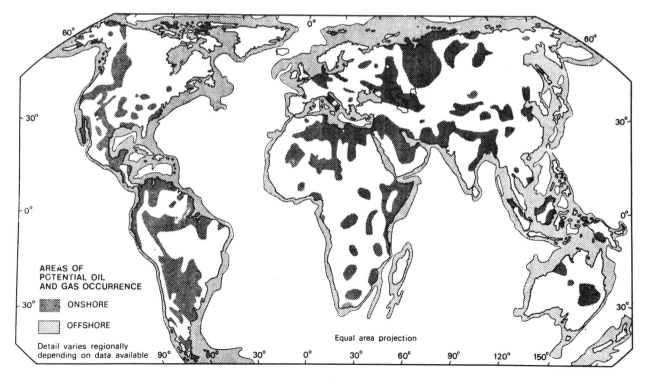

AREAS OF POTENTIAL OIL AND GAS OCCURRENCE

ONSHORE

OFFSHORE

Detail varies regionally depending on data available

Equal area projection

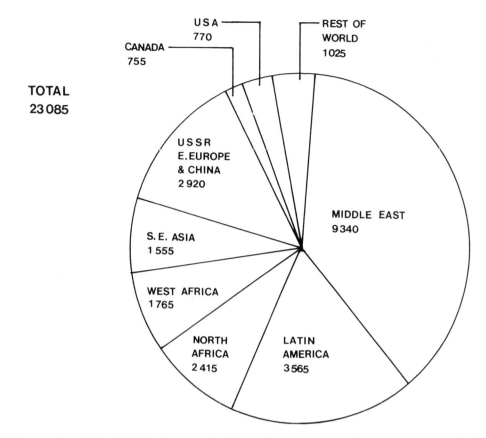

USA
770

CANADA
755

REST OF
WORLD
1025

TOTAL
23 085

USSR
E. EUROPE
& CHINA
2 920

MIDDLE EAST
9 340

S. E. ASIA
1 555

WEST AFRICA
1 765

NORTH
AFRICA
2 415

LATIN
AMERICA
3 565

World exports of oil in 1985 (in thousands of barrels daily).

Most of the major oil-producing countries are also major exporting countries too. This is particularly true of the Middle Eastern states, most of whom have very limited need for oil themselves. The Middle East is the richest oil region in the world by far. Its oil has the advantage of occurring in very large "pools" (these are not underground lakes as the word might suggest, but concentrations of oil collected in the pores of oil-bearing rocks). These pools are near to the surface and the oil trapped in them is also subject to high pressure so that it doesn't need much pumping out of the ground. It is therefore very cheap to extract. The Middle East oilfields are located round a number of shallow sea gulfs of which the biggest by far is the Persian (or Arabian) Gulf. Oil is also found in Iraq and Libya.

Although the Middle East as a whole only produces about as much oil per year as the Soviet Union, it is the world's largest *exporter* of oil, accounting for nearly 40 per cent of the world's

exports in 1984. However, it is a much less important exporting region than it was 20 years ago. This is because political conflict in the region has made importing countries, particularly the USA, more cautious about becoming dependent on the Middle East for their oil.

The USA today buys much of its oil from Latin America while Japan and Australia are relying more on supplies from South-East Asia. Nigeria, Mexico and the North Sea have all now become major competitors of the Middle East in the oil business. Nevertheless, the reserves of Middle East oil still exceed those of all the rest of the world put together. Nearly 60 per cent of the world's reserves are found there and nearly half of these are in Saudi Arabia alone. At present rates of production (in 1984) there is enough oil in the Middle East to last over 100 years, ten times longer than the current American reserves and five times longer than those in the North Sea. Middle Eastern oil is also by far the cheapest in the world, costing a mere $3-5 per barrel to produce compared with anything from $5 to $40 a barrel in the North Sea.

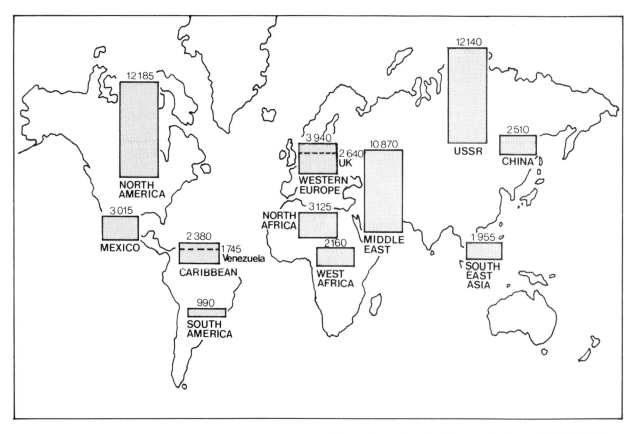

The main centres for the production of oil in 1985 (in thousands of barrels daily).

Low-lying sea gulfs, deltas and river basins elsewhere have proved to contain oil-bearing rocks. The Gulf of Mexico is a major production area, and so too are the Maracaibo basin in northern Venezuela and the Niger river delta in Nigeria. Much oil today is found in offshore fields, however. New discoveries have been made in recent years off the coasts of Sarawak, Brunei and Malaysia and these, when added to Indonesia's output from Sumatra, make South-East Asia a significant oil producing region. The biggest offshore finds of all were made in the North Sea in the 1970s, following the discovery of the natural gas field of Gröningen off the Dutch coast in 1959. In 1969 the Ekofisk oilfield was found off the coast of western Norway, and then, in December 1970, the huge Forties field, one of the largest oilfields in the world, and the first of many to be discovered in subsequent years. By 1985 Britain was the world's fourth largest oil producer.

* * *

The biggest consumers of oil are the advanced industrial nations. Not only is car ownership in these countries the highest, but so also is the demand for oil

from industry. Many of these richer countries are also located in the cool temperate parts of the world where houses and offices need to be heated in winter. The extent to which different countries rely on imports, however, varies considerably. Since the Second World War the USA has shifted from self-reliance in oil to import-dependence, while Britain has gone in the opposite direction. Japan, with little coal, is almost completely dependent on oil imports for her energy needs.

Energy consumption per head of population also varies widely within the rich countries. The Americans are by far the most prolific consumers of oil, due, many would say, to their extravagant and wasteful use of it. With only just 6 per cent of the world's population they consume no less than 29 per cent of the world's oil, something of a scandal when you think that wood is still the main fuel for 80 per cent of the people who live in poor countries! Even in the poor countries the demand for oil has gone up, despite the considerable rise in its price between 1973 and 1980. These nations today account for nearly 25

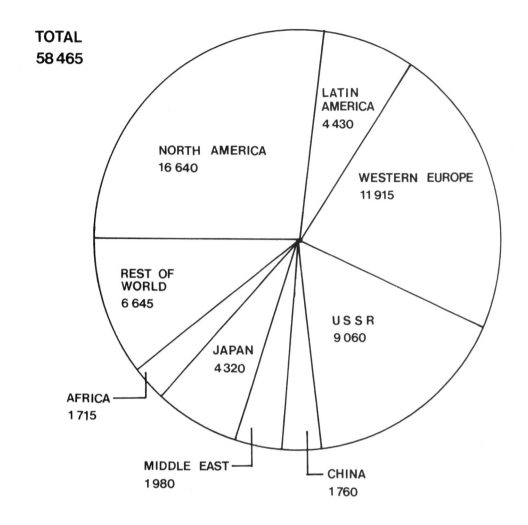

TOTAL
58 465

LATIN AMERICA
4 430

NORTH AMERICA
16 640

WESTERN EUROPE
11 915

REST OF WORLD
6 645

USSR
9 060

JAPAN
4 320

AFRICA
1 715

MIDDLE EAST
1 980

CHINA
1 760

World consumption of oil in 1985 (in thousands of barrels daily)

per cent of the world's oil imports. This is because their populations are rising very rapidly and because a small number of them like South Korea, Taiwan and Nigeria are also industrializing at a fast rate. The price rises between 1973 and 1980 cut world oil consumption for a time and certainly forced the rich countries to use oil less wastefully, but with the fall in oil prices in the 1980s consumption has already crept back to its 1979 levels. This shows how dependent the world has become on oil.

Since most of the major consuming countries lie some distance away from where oil is produced, a large part of the world's oil has to be moved long distances to where its markets are. Oil can be transported either by pipeline or tanker. In remote and virtually uninhabited areas, like deserts, pipelines can be laid above the ground. Elsewhere, however, they must be buried below ground for safety, technical and environmental reasons. Before laying the pipe, the route has to be carefully planned to avoid natural hazards like rivers and high ground. Rights of way may have to be purchased and the route has to satisfy local authorities and environmentalists. Initially, a deep trench is dug and lengths of pipe are welded together before they are lowered into the trench. The pipe is then coated with bituminous enamel or plastic tape to protect it. In Alaska the ground is frozen too hard for pipes to be buried and they have to travel above ground. Several different refined products are often transported in the same pipeline. Special pumping stations are used at various points along the pipeline to boost the pressure in the pipe: by increasing the pressure the flow of oil is increased.

Laying underwater pipelines is more difficult and also more expensive. There, the pipe has to be coated with reinforced concrete so that it will sink below the surface and stay there. It is laid by a special pipe-laying barge. The lengths of pipe are welded together on board and then fed out over the stern of the vessel. In shipping lanes the pipeline has to be laid on the sea bed in order to avoid getting tangled up with passing

ships. Pipelines are used to transport both oil and natural gas. They can be very long. Much of Russia's oil is brought thousands of miles across the country by pipeline from Siberia, and there are also plans to pipe natural gas all the way from the Soviet Union to western Europe.

Eighty per cent of the world's oil today, however, is moved by tanker, and oil tankers make up a large part of the world's merchant fleet. Only about a third of these tankers belong to the oil companies; the rest are chartered from large ship-owners like the Onassis family. Most tankers ply the world for hire, a bit like taxis. The ship's master will receive his instructions from the owners or the charterers over the radio telephone, telling him where he is to go to pick up a delivery and where it is to be delivered. In 1984 there were over 3000 tankers in existence, including 71 supertankers, giant ships of over 320,000 tons deadweight, some of them as long as four full-sized football pitches placed end-to-end! A tanker this size needs at least 60 feet (18½m) of water to load and unload its cargo. Few harbours in the world possess this depth of water and therefore special deep-water oil berths have had to be constructed to take them. You can see one of these at Bantry Bay off the south coast of Ireland and they are very common in the Persian Gulf.

Large tankers run the risk of causing pollution if they collide at sea or go aground. In 1967 the Liberian tanker *Torrey Canyon* broke in two after striking the Seven Stones Reef off the Isles of Scilly, spilling into the sea over half of its 119,000 tons of oil. As the vessel broke up it was actually bombed by the Royal Air Force in an attempt to burn up the oil which was spreading over the sea in a huge slick. The worst accident of all occurred on 16 March 1978 when the 232,000-ton tanker *Amoco Cadiz*, fully loaded from Kharg Island in the Persian Gulf to Rotterdam, went aground on the Brittany coast after its steering gear failed in rough seas. Its oil polluted many miles of the French coastline, killing off fish and covering thousands of diving birds with oil.

The world's most important tanker routes today are the western Indian Ocean route round the Cape of Good Hope, which takes most of the bulk oil from the Middle East to Europe, and the eastern Indian Ocean route from the Persian Gulf to Japan. There is also a sizeable oil traffic along the eastern seaboard of the USA. The most heavily used sea lane in the world is the Strait of Hormuz at the entrance to the Persian

The main overland pipeline from the Fahid oilfield in Oman to the Mina-al-Fahal oil terminal on the Gulf of Oman. In the background are the oil storage tanks from which ocean-going tankers are loaded with oil. The oil is piped aboard under high pressure.

A frogman links up two sections of BP's undersea oil pipeline from the Forties field to the Scottish coast. This picture was taken 65 miles out in the North Sea.

42

One of the giant "supertankers" built in the 1970s to carry oil from the Middle East to Europe and Japan. This is the 553-tonne *Batillus* owned by Shell of France.

The large modern oil terminal at Mina-al-Ahmadi, Kuwait, on the Persian Gulf.

Gulf. Two-thirds of the world's sea-borne trade in oil (80 ships a day) passes through it. Its waters are treacherous, with strong currents, low swells and many small islands between which large tankers have to navigate. There is also very poor visibility. The Iran-Iraq War which broke out in 1980 has led to the additional hazards of mines in the sea lanes and rocket attacks on shipping.

Given the huge changes that have occurred in the oil industry it is not perhaps surprising that the geographical pattern of oil refining has changed over the years. In the 1950s much oil was refined at or near the place where it came out of the ground. That often meant the Persian Gulf. This was because a large weight-loss could be effected by refining the oil before it was shipped across the sea by tanker. Changing refinery techniques, plus the growing political instability of the Middle East, have changed all that. Today, weight-loss from refining on the spot is much less and many countries have tried to reduce their dependence on the Middle East for political reasons.

As the price of oil rose in the 1970s it also became cheaper for most countries to import crude oil and

The stricken oil tanker *Torrey Canyon* as she broke up on the Seven Stones Reef, off the Isles of Scilly, in March 1967. Oil spillages resulting from wrecks like this threaten coastlines and marine life for many miles around.

refine it themselves. Many large oil refineries were therefore built on the coasts and river estuaries of north-western Europe and Japan. Today, the Scheldt estuary with its great complex of oil installations around Antwerp, Rotterdam and Amsterdam, and the Ellesmere Port-Stanlow area of Merseyside have become major centres of oil refining. So have Marseilles on the Rhône delta in southern France and the Yokohama-Kawasaki region on Tokyo Bay in Japan. The Mediterranean ports of southern Europe have in addition become important trans-shipment centres for oil pipeline networks, which carry oil deep into the industrial heartlands of western Europe.

THE POLITICS OF OIL

Many people were surprised that the OPEC countries were able to cooperate so well in the 1970s in fixing prices and voluntarily restricting their production of oil. They seemed at first sight to have little in common; they were geographically dispersed, with very different histories, social systems and forms of government. Yet there were many things that cut across these differences of interest and background. First of all, they were all poor countries or countries that had recently been poor. Many of them, like Nigeria, Iraq and Libya, also had cause to resent their former colonial status; while the South American states, even though they had long since ceased to be colonies in the formal sense, lived nevertheless under what they also saw as the semi-colonial exploitation of the Western oil companies.

A majority of them – 11 out of 13 – shared a common religion too: Islam. In addition to the Middle Eastern members of OPEC, countries as geographically distant as Nigeria and Indonesia were also states with Islamic majorities. This gave them a loyalty to each other as co-religionists, and also a sense of being part of a religious movement which had historically challenged the domination of the Christian West. Eight out of the 13, including the largest oil producers, were located in the Middle East, and were therefore climatically and culturally similar to each other. There were of course important economic differences between them, particularly between the countries with large urbanized populations (Iran, Iraq) and the city states of the Arabian gulf with their small, dispersed pastoral populations but few other resources. These differences, nonetheless, were not important enough to cause any breach in the ranks of OPEC during the prosperous days when oil prices were rising (things were to be very different later, however, when the demand for OPEC oil began to drop sharply in the 1980s).

Most importantly, seven of the eight Middle Eastern states were also Arab countries; that is, they spoke Arabic and saw themselves as an integral part of a historical Arab civilization. (The odd country out in the Middle East was Iran. Although Persia – the old name for Iran – was also an ancient society, the Iranians, despite being Muslims, are not Arabs; their native language is Persian or Farsi, not Arabic, though they do use the Arabic script.) The fact that the most important members of OPEC were Middle Eastern Arab countries had profound political consequences. It meant inevitably that the question of oil prices and oil production would get caught up in the complex web of Middle Eastern politics. At the centre of that web was the long-standing and deep hostility of the Arab nations to the state of Israel.

Israel had been created in 1948. Its creation had involved the violent displacement of many Arabs from their homeland, the old province of the Turkish Empire known as Palestine. These Arabs became refugees and some of them were forced to live in refugee camps in various parts of the Arab world. This event was universally seen as a defeat and a humiliation by the whole Arab world. To the Jews, however, many of whom had emigrated to Palestine both before and after 1948, Palestine (or Israel as it now became) was the Biblical home of their forefathers and a country to which they had a historical claim. The Arabs, however, believed that Israel would never have come into existence but for Britain and the USA. In 1917 the British Foreign Secretary, A.J. Balfour had promised the support of the British government for the establishment of a Jewish national home in Palestine. After the First World War the British government was given responsibility for Palestine under the League of Nations mandate and for a few years permitted Jewish immigration into the country. It was in the USA, however, where there was a large Jewish community, that the support for Zionism (the belief in the right of the Jews to their own homeland in Palestine) was strongest. After the Second World

War, and under pressure from the Zionist lobby, the US government played a leading part in bringing about the creation of the new Israeli state. Not surprisingly, therefore, the Arabs regarded Britain and the USA with some hostility.

The Arabs were unable to destroy Israel militarily, however, mainly because she was able to rely on American backing and arms. They fought – and lost – three wars against the Israelis (in 1948, 1956 and 1967). By 1973 they had come to realize that Israel could only be dislodged, or at least made to negotiate with the Arabs, if the Americans and other Western countries could be forced to modify or even abandon their support for Israel. Another Arab-Israeli war broke out in October 1973 during the Jewish passover festival of Yom Kippur. This time, however, the Arabs had, as they thought, a weapon

they hadn't possessed previously – the "oil weapon". As oil supplies from the Middle East dried up on the outbreak of war, the Arab countries within OPEC tried to impose an oil embargo on those Western countries, including the USA , who were particularly pro-Israel. They threatened to withhold supplies of oil to these countries unless they gave up their unconditional support for Israel.

In certain respects the boycott was not a success. It was impossible to prevent the USA from obtaining the oil she needed from other parts of the world, particularly the Caribbean. Indeed, she was aided considerably in this by the big oil companies, five of whom were American-owned. The only way an oil embargo could have been made to work would have been to impose it on *all* the Western oil consumers. There was no support for this, however, within OPEC since it meant damaging pro-Arab as well as pro-Israeli countries. It also confronted the OPEC members with what was to be a permanent dilemma for them – just how far could they go in inflicting damage on countries on whom they depended to sell their oil?

After 1973 the oil-producing countries were no longer under the control of the Western oil companies. In this cartoon, the genie refuses to be the slave of the oil tycoon who rubs the magic oil lamp (see the story of Aladdin and His Magic Lamp in *The Thousand and One Nights*).

The Persian prime minister, Dr Mohammed Mossadeq, is mobbed by enthusiastic crowds as he rouses anti- **British frenzy against the Anglo-Iranian Oil Company, Teheran 1951.**

In the end OPEC realized that it could not afford to undermine the prosperity of the Western countries *too* much. Ironically, the attempt to use oil as a political weapon probably worked in the end to the disadvantage of the Middle East, since from 1973 onwards many oil-consuming countries started to look for oil permanently in other parts of the world where the supply would not be disrupted by political conflict. However, there is no doubt that it did bring some political benefits to the Arabs, since a number of west European countries, realizing that it would be unwise to alienate the Arab oil producers, moved closer towards recognizing the claims of the Palestine Liberation Organization, which wanted an Arab state re-established in Palestine.

* * *

Israel was not the only issue in Middle East politics. Although all the Arab countries were united in opposition to Israel they were also divided among themselves. One division was particularly important – that between the radical and the conservative Arab states. These states each used the power that their oil wealth gave them to further their own political aims and objectives in the Middle East, and the aims of one state were often in sharp conflict with those of another. Oil thus contributed to the sharpening of internal political tensions in the Middle East.

The radical states were led by Libya, whose

Colonel Muammar Qadhafi, President of Libya, addressing a meeting of his supporters.

Iranian troops dig in during the Ramadhan offensive in
the "Gulf War" against Iraq, July 1982.

president was Colonel Muammar Qadhafi. Libya, like neighbouring Algeria, had been a former Western colony before it became independent in 1951. Qadhafi was bitterly anti-Western and saw Western powers like Britain, France and the USA as capitalist, imperialist states who had oppressed poor states like his own and who, given the chance, would do so again, this time through the power of their oil companies. He therefore wanted to regain control of Libya's oil supplies in order to reduce the volume of oil exports and thus get a better price for the oil. He was also willing if necessary to deprive the West of some of its oil supplies.

Qadhafi also had ambitions to be the leading figure in the Arab world. He had already won great prestige as the man who had taken on and destroyed the power of the oil companies in 1969-70. In political matters he was a radical who modelled himself on his hero, Abdul Gamel Nasser, the president of Egypt from 1956 until 1970. Nasser had won the hearts of the Arab masses by expelling the British from Egypt in 1956 and transferring the wealth of the rich Egyptian land-owning class into the hands of the poor peasants (the *felaheen*). Qadhafi had similar plans for Libya, and urged radical groups in other Arab countries to overthrow their ruling classes as well. He now used Libya's new oil wealth to build up one of the largest and most modern armed forces in the Middle East, and started to finance pro-Qadhafi groups in other countries.

The conservative states were led by Saudi Arabia and included most of the sheikhdoms of the Persian Gulf as well as a number of non-oil states like Jordan and Morocco. Saudi Arabia and the Gulf states were all controlled by tribal dynasties. The rulers of the Gulf states had been brought under the protection of the British government in the nineteenth century as a way of preserving British influence over the strategically important imperial trade route to India. Not surprisingly, Saudi Arabia and the rulers of the Gulf states didn't share many of the anti-Western attitudes of Qadhafi and the other radicals in the Middle East, like Iraq and Syria. They also needed the support of the West. It was in the rich economies of Europe and North America that their huge oil revenues were invested. Many of them also owned property in the West and came to Europe for medical treatment; their sons, too, were often Western-educated. They also needed highly sophisticated

modern weapons for their armed forces; for these they looked to the West. Not surprisingly, therefore, they had an interest in keeping down the price of oil to a level which didn't cripple the Western economies. However, as Arab countries, they came under pressure from their Arab neighbours in 1973, and after, to use their oil power against the West as a means of forcing the Western powers to abandon their support for Israel. Saudi Arabia, in particular, both as the leading oil producer in the Middle East and as the leading upholder of Muslim orthodoxy and guardian of the holy city of Mecca, has had to follow an uncertain middle course between these two conflicting sets of pressures.

Politics in the Middle East have also been shaken by another development which has had profound consequences for the international oil industry. Although Islam has been a unifying factor in the Muslim world, it has only united Muslims against non-Muslims. Within the Muslim world new conflicts have arisen in the 1980s, some of them the product of the changes which the oil revolution has wrought on Middle Eastern society. While the new oil wealth made possible, in theory, huge advances in living standards in poor countries like Iran and Iraq, in practice it often went to making the rich in those countries richer still. There remained many of the poor – perhaps even the great mass – who resented this deeply. The poor, therefore, increasingly turned to radical political leaders, some of whom, like their *mullahs* (or religious leaders), were deeply anti-Western. In the more conservative countries there was also a fear that old Islamic traditions were being corrupted by the growth of Western materialism. These forces and fears led to the revival of old conflicts within Islam. These tensions reached their peak in Iran in 1979 when the pro-Western Shah was deposed and replaced by a militant, anti-Western régime led by one of the leading *mullahs*, the Ayatollah Khomeini. In 1980 war broke out between Iran and neighbouring Iraq. Both these states were major oil producers. The revolution in Iran brought about a collapse in Iran's oil exports and the subsequent war had a similar effect on Iraq's too. This collapse in the supply of oil to world markets led during 1979-80 to a four-fold increase in its price. This had catastrophic consequences for the oil-consuming countries of the world, rich and poor alike. It demonstrates as vividly as anything can do

The Saudi Arabian delegation at the Geneva meeting of the Organization of Petroleum Exporting States, 27 December 1984. Sheikh Yamani, the Saudi oil minister, is second from the right.

the ways in which the world of politics and the world of oil are indissolubly linked.

* * *

Today the OPEC states are in disarray. As the high price of oil forced down the demand in the early 1980s, so the members of OPEC were under pressure to cut back their output in the hope of staving off a total collapse of prices. The old unity, regional and religious, which sustained OPEC in the years of rising prices gave way to a new disunity as a growing conflict emerged between those countries like Nigeria, Ecuador and Venezuela who, for domestic economic reasons (large populations and, therefore, high dependence on imports), sought to sell as much oil as possible before prices collapsed even further

Newspaper headlines show how large fluctuations in oil production and prices affected the oil-producing countries.

NO WONDER SHE'S LAUGHING. SHE'S GOT SCOTLAND'S OIL.

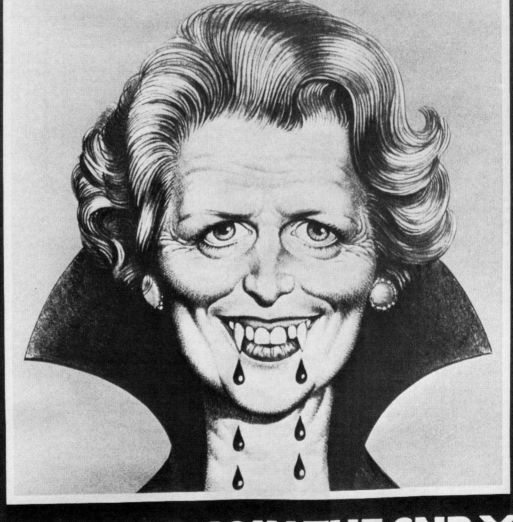

STOP HER. JOIN THE SNP.

and those countries who were under no such pressure, like Saudi Arabia. Ironically, among those countries pressing hardest to sell more oil were Iran and Iraq, who needed the revenues from their oil sales to finance their costly war with each other!

Has oil raised the level of *world* tension? So far, the major world powers, the USA and the Soviet Union, have managed to avoid getting caught up in a situation where their armed forces confront each other over the Middle East, though the USA has always been ready to protect Israel indirectly by supplying her with arms whenever the Arabs threatened her. There *was* a time certainly, in the 1950s, when the USA, Britain and France feared that the Soviet Union would try to take advantage of their dependence on Middle East oil. They feared that Russia would try to draw the Arab states into the Soviet sphere of influence and cut the West off from its oil supplies. This did not happen, however. Partly, this was because the Soviet Union was not herself in direct competition with the USA for scarce supplies of Middle Eastern oil. She has her own oil and, anyway, has always tried to avoid becoming dependent on outside supplies. As a small oil exporter today she is in some ways a commercial rival of the oil-producing countries of the Middle East. But, above all, she has feared getting involved in an armed conflict with the West which

A Scottish Nationalist Party election poster (1983). The Scottish Nationalist Party made an election issue out of the British government's claim that North Sea oil belonged to the United Kingdom as a whole and not just to Scotland.

could easily escalate into a nuclear war. Today, she is anxious to sell natural gas to the West in return for scarce Western currency and imports of Western technology. Unless she ceased to be able to supply her own oil needs and was forced to buy oil from the Middle East, there seems no great likelihood of a major confrontation between the Soviet Union and the West, at least over the question of oil supplies.

One country which is crucially dependent on outside supplies of oil is South Africa. Faced as she has been for a number of years with the threat of economic isolation from the rest of the world because of her apartheid policies, she has been forced to turn to manufacturing oil synthetically from coal. She has also built up a considerable stockpile of imported oil. Despite her reserves of coal, however, only about 40 per cent of her oil needs can be met by the SASOL process of producing oil synthetically from coal. If comprehensive economic sanctions are applied against South Africa her economy could be seriously damaged, perhaps bringing nearer the day when the apartheid régime is overthrown. Much, however, will depend on whether sanctions can be enforced against her. South Africa's future, whether under her present government or a black majority government, concerns, or should concern, all the Western oil-importing countries, since the great bulk of the world's sea-borne oil traffic now goes round the Cape of Good Hope. In any possible future conflict between pro- and anti-Western interests in southern Africa the power to determine the fate of shipping rounding the Cape will be of crucial importance.

THE FUTURE OF OIL

The 1973 rise in oil prices caused the world demand for oil to level off, while the 1979 increases actually caused it to fall. Between 1979 and 1982 consumption fell by 20 per cent. Most of the immediate reduction was due to economies made in the use of oil by the advanced industrial countries. Consumption of oil per head fell by 19 per cent in the USA and 22 per cent in Japan during that period.

Oil can be conserved in a number of ways – by

A big "gas-guzzling" American car of the 1950s. Cheap petrol encouraged the production of large cars with high petrol consumption. Big cars also had sex appeal!

countries going over to the production of smaller cars that use petrol more economically for instance. Before 1973 the Americans were the worst offenders in the extravagant use of gasoline. In that year the "gas-guzzling" American fleet burned half the world's petrol! Petrol can also be saved if substitutes can be found for it. In 1983 some 25 per cent of Brazil's petroleum needs were being met by alcohol distilled from sugar cane. A second way of saving fuel is by reducing heat loss from houses, offices and public buildings. This can be done through more efficient insulation and by turning down central

heating (or air conditioners in hot or humid climates).

There have always been alternative sources of energy to oil. Whether these will be used in the future depends on two things: the price of oil compared to the price of alternatives, and whether these alternative fuels can match the advantages which oil unquestionably has as a source of energy. The future price of oil can only be guessed at. It may continue to fall if the OPEC countries fail to recapture control of the world's oil markets. In the longer term, however, there seems to be a good chance it will rise, since the demand for oil in the newly industrializing parts of the Third World is rising fast. Even in the countries which are industrializing less rapidly the demand for kerosene as a cooking and heating fuel is certain to go up as populations increase. We can only guess, too, at whether new oilfields will come into production, particularly in countries like China, where there may well be untapped reserves of oil that could be exploited cheaply.

It is difficult to predict with accuracy what the levels of the world's reserves will be in the future. The oil companies keep their own estimates a closely guarded secret, but if new low-cost recovery techniques are developed by the oil industry, much oil that at present is too expensive to get out of the ground could become economic to recover. Given existing levels of oil-recovery technology, however, "proven" or recoverable oil reserves at present amount to about 650 million barrels, with the possibility that there are another 600 million barrels as yet undiscovered. This amounts to two and a half times all the oil consumed since 1859. At the levels of demand prevailing in the early 1980s those reserves could be used up early in the twenty-first century. If we are wise, therefore, we will plan for a world in which oil will be a permanently scarce resource.

In 1984 oil supplied 44 per cent of the world's commercial energy. It has many advantages over alternative sources of fuel. Compared to other fuels like coal it is clean to use and relatively easy to transport; it can also be used for a variety of different purposes, not just as a fuel but also as a "feedstock" – that is, a chemical source for many other by-products like artificial fibres, detergents, polishes and plastics.

World oil reserves (in thousand millions of barrels currently proved), 1985.

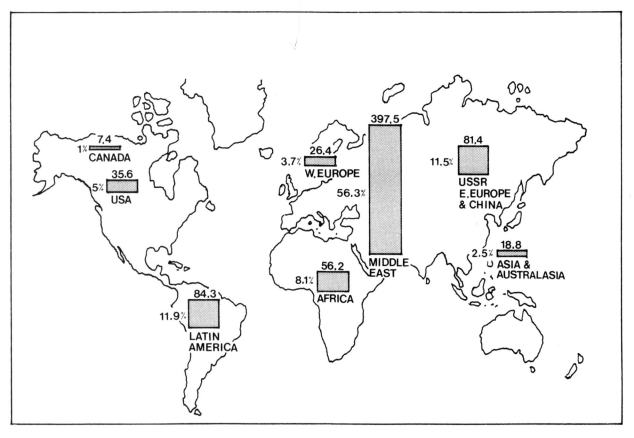

What are the alternatives to oil? First of all there are other fossil fuels. Fossil fuels are fuels that come from fossilized animal and vegetable matter found in the geological strata of the earth's surface. Natural gas is one. It is found together with oil and today supplies 20 per cent of the world's commercial energy. But most of it comes from just five countries – the Soviet Union, Mexico, Iran, the Netherlands and Algeria. It is also less easy to transport than oil. For transportation over long distances across the sea it has to be liquified and kept at very low temperatures in special high-pressure container ships. As yet there are few undersea pipelines, though several are planned for the purpose of bringing natural gas from North Africa to the Mediterranean coasts of France and Italy and one has already been completed. The Russians are also anxious to build an overland gas pipeline from Siberia to western Europe. The use of natural gas, however, is limited to domestic central heating and cooking and even then it can be used only in densely populated areas which can be easily linked to gas pipeline networks. Rural areas may have to remain dependent on kerosene for a long time to come, if not for ever. Natural gas also has a limited

range of uses: it can't be used at present as a fuel for motor vehicles or as a chemical feedstock for instance.

Coal supplied 27 per cent of the world's commercial energy in 1984. Its main advantage is that there is plenty of it. It is reckoned that recoverable reserves of coal will last another 270 years. However, most of it is produced by just ten countries; 57 per cent of the world's reserves today are found in the USA, the Soviet Union and China, although Poland, Australia and South Africa are substantial producers, as is India. Other problems with coal are that its use pollutes the atmosphere (all those chimneys!) and degrades the landscape with spoil-heaps. Acid rain, produced when the sulphur and nitrogen dioxide from coal fumes and smoke combine with water in the atmosphere, can also have a disastrous effect on natural vegetation. Moreover, coal mining is a dangerous job – over 2000 miners a year lose their lives in the coal mines. To make matters worse, coal is bulky, slow and expensive to transport round the world. It also has a far narrower range of uses than oil, being largely limited to providing power station fuel and domestic heating.

Of the non-fossil fuels, nuclear power supplied just two per cent of the world's energy in 1984. Like coal, its use is largely limited to powering electricity generators but it suffers today from a number of other major drawbacks. The nuclear disasters at Three Mile Island in the USA in 1979 and Chernobyl in the Soviet Union in 1986, when radioactivity leaked from crippled nuclear reactors, have raised public fears about whether nuclear energy is safe. There is also public disquiet about whether nuclear waste can be safely disposed of, and many people are worried as well that governments with a nuclear energy industry can far too easily switch it over to the manufacture of nuclear weapons. There are now powerful anti-nuclear movements in western Europe and Japan. Of the major industrial powers only France is pressing ahead with a nuclear power programme today; by contrast, many others like Sweden and Austria are phasing their programmes out altogether. The future of nuclear energy as a major energy source, despite its huge potential, seems at present to be in some doubt.

What about renewable energy resources – those which are not destroyed in the process of energy-creation? Water power (hydro-electric power) can make a contribution to energy needs in mountainous countries where lakes and rivers can be easily dammed. In those parts of the world where its potential is greatest, the Third World, it would only contribute a small percentage of total energy needs, however. The cost of building dams and power stations is very high anyway, and probably beyond the means of most poor countries, certainly if they have no outside assistance.

Polluted air was one of the results of using coal-fired steam engines to drive factory machinery in the earlier part of this century. This was the scene in a Lancashire cotton town in 1931.

The crippled nuclear power station at Chernobyl,
Ukraine, USSR, May 1986. The nuclear fallout following
the Chernobyl disaster reinforced public fears about the
safety of nuclear power.

Police removing demonstrators protesting at the
building of a new nuclear power station at Brokdorf,
West Germany, 1981.

Solar energy – heating water or cooking food using the direct rays of the sun or generating electricity by the method of directing solar rays on to energy-storing solar panels or cells – could make a minor contribution to the energy needs of warm countries. Israel is a good example. It has a limited use, however, in the northern industrial countries of the world, where the demand for energy is greatest. If we exclude the wood that is used for heating and cooking in many parts of the Third World, it seems that oil will remain the single most important, if not the sole, source of energy in many parts of the world for a long time to come.

*   *   *

As the advanced countries of the world move into a more oil-intensive economy they will have to come to terms with many new problems. Despite being cleaner to use than some other fuels, the scale on which oil is used today already makes oil pollution a problem. Atmospheric pollution occurs through emissions into the air of sulphur dioxide, nitrogen oxide and carbon monoxide – all unburnt hydrocarbons. These come from factory chimneys, chemical works, power plants and especially car exhausts. We have already seen the damage done to the marine environment by oil spillages and waste discharges. The excessive use of petrochemicals like fertilizers, pesticides and detergents has also been shown to have serious effects on animal and plant life.

Many threats to the environment can be removed by imposing tougher penalties on polluters, or changing the design of polluting technologies. It can be made an offence to discharge oil waste into river

Solar power – the use of the sun's rays to generate electricity. The picture shows a solar power unit in the desert near Albuquerque, New Mexico, USA. Notice the solar reflector discs in front of the tower.

estuaries or emit chemical fumes into the atmosphere. Cars, too, can be adapted for lead-free petrol, though designing and producing cars with pollution-free exhaust systems is expensive and puts up the cost of manufacturing a car – and therefore its price. Lead-free petrol is also more costly to produce than ordinary petrol. We have, therefore, to balance the advantages of a pollution-free atmosphere against the extra costs of bringing it about, though many people would still think the extra cost was worthwhile.

If we *really* want to avoid polluting our environment we would do well to recognize that our whole way of life is much to blame, particularly our heavy dependence on the motor car. There are over 300 million cars in existence today, well over half of them in the advanced industrial countries of the

world. Car ownership has gone furthest in the USA, where the average family now has two cars. There, the "car culture" has developed a firm hold. The status of people is rated by how many cars they have, and how big and powerful these are. The USA in the 1940s and 1950s was the home of the big "gas-guzzling" automobile with its very high petrol consumption and low mileage per gallon. This

fashion for big cars spread to Europe in the 1950s, particularly through the influence of Hollywood films. More and more cars were styled along American lines, and big American cars were made to appear glamorous, along with the American way of life generally.

Bigger cars and more widespread car ownership in the USA meant there was a demand for more and bigger motorways (called "freeways" in the USA), not just inter-city motorways either, but also freeways within the cities themselves. Today, 30 per cent of all the land in the Greater Los Angeles area is accounted for by freeways, and Los Angeles itself has developed into a huge suburban sprawl as residential housing estates have mushroomed out from the freeway system. It is even said that to be seen walking in the suburban streets of Los Angeles makes a person an object of suspicion to the traffic police! In the morning and evening rush hours, cars built to travel at 90 mph crawl bumper-to-bumper along the free-way system at 12 mph. This same pattern of traffic congestion is repeated in Tokyo, Lagos, Bangkok and many other cities around the world.

Along with traffic congestion goes the problem of pollution from car exhaust fumes. Los Angeles and Tokyo suffer from some of the world's most appalling photochemical smogs. There are also 50,000 highway deaths per year in the USA. The "car culture" of course also penalizes those too poor to own cars. If they live on the outskirts of the city they find themselves having to depend on inadequate public transport – inadequate because most people have cars and therefore little interest in an adequate public transport system. Alternatively, they have to pay high rents if they are forced to live near their work in the city centres – because of not having a car!

Since the oil price rises of the early 1970s, however, even the Americans have had to start thinking of ways of economizing in the use of petrol. The "love affair with the motor car" seems to be over today, or at least on the wane, as more American cities are rebuilding cheap and economical "mass-transit" systems, forms of public transport which enable people to move quickly by electric train from the suburbs to their places of work in city centres. Elsewhere, too, there

Rush-hour traffic in Bangkok, Thailand. By the 1970s the boom in car ownership was making traffic congestion a major problem in most of the world's large cities.

have been moves towards ways of living that make us less dependent rather than more dependent on petrol.

* * *

If the world is going to rely on oil as its single most important source of fuel in the future another problem we shall have to face is just how the available supplies are going to be shared out. Will the rich countries be able to outbid the poor countries for what supplies of oil are going, thus perpetuating their poverty and holding back their efforts to develop? Or will the rich forego some of the uses to which they put oil at present, to allow the poor countries a larger share? Oil is important to most manufacturing industries today. If we were honest with ourselves we would probably admit that our need for higher and higher levels of production in the advanced countries is less urgent than that of poor countries, where

people often lack some of the basic essentials that we have long taken for granted – things like bicycles, radios and kitchen equipment. Oil there is also needed for irrigation pumps and village electricity generators, while pharmaceuticals and fertilizers are required for improving health and the quality of crops.

Only by foregoing a part of what we keep for ourselves at present can we help other, less fortunate, countries. But many things get in the way of our doing this. In the rich countries, people are encouraged by advertisers to consume more and to crave for the latest, most fashionable products, whether they are really necessary to our lives or not. (Do we really need electrically operated wind-down car windows, for instance?) People are easily persuaded that possessing more consumer goods means they are "keeping up with the Jones's". It is not easy to convince them that a higher standard of living doesn't just depend on having consumer goods but can also result from living in cleaner, less crowded cities, suffering less stress at work and having more leisure time.

Most of us have little first-hand knowledge of just how squalid the living conditions are in many Third World countries. Perhaps if we did we would be readier to make sacrifices to help people there. When scenes of terrible starvation were shown on British television during the Ethiopian famine of 1985, millions of individuals sent in donations to the Famine Appeal Fund; the same thing happened in 1986 when Bob Geldof organized pop concerts to aid the poor and starving in Africa.

Whether the crisis point is reached where there just isn't enough oil to keep factories running, cars moving and homes heated depends partly on how fast the demand for oil rises in the next 20 years and on how quickly new reserves can be tapped. If demand rose at the speed it was rising at the beginning of the century (just about doubling each year) we should be experiencing what is called an *exponential rate of growth* of demand. To appreciate just how explosive an exponential growth rate could be, it is worth looking at the famous example of the lily pond. Imagine a pond where the number of leaves floating on the surface of the water doubles each day, so that there are two leaves on the second day, four on the third, and so on. The pond is completely full of leaves on the thirtieth day. But it was only half-full on the twenty-ninth day! Such a rate of increase in the demand for oil would quickly lead to a crisis, with oil supplies running out very quickly indeed.

Today, fortunately, the demand for oil is well

The Gravelly Hill interchange, West Midlands, UK, otherwise known as "Spaghetti Junction". Motorway complexes like this mushroomed in the 1960s and 1970s with the huge expansion in private car ownership in Britain and western Europe.

below the exponential level, but if pre-1973 levels of demand were to revive, existing reserves would still be exhausted within 30 to 40 years, leaving future generations without any oil at all. At present levels of consumption there is enough oil in the ground to last for perhaps two more generations. Much will depend on how far the oil companies themselves drill for oil. New reserves will be tapped only if prices rise high enough to make oil exploration profitable. Improved techniques for recovering oil will of course make it more attractive to drill for oil in areas which are at present unprofitable. Political considerations, however, can also influence oil company policies

"Smog" (the pollution of the atmosphere by, among other things, car exhausts) is becoming a major environmental hazard. A cyclist wears a protective mask while cycling through traffic in Washington, DC, USA.

here. A number of Third World countries who at present import oil are believed to have substantial reserves of untapped oil (Jamaica is an example) but the oil companies have in the past avoided surveying or prospecting in countries whose governments they suspect might nationalize foreign-owned oil wells. The extent to which reserves are drawn upon, therefore, is only partly an economic issue.

The future of oil as a source of energy is fraught with many uncertainties. Improved oil recovery techniques, changes in lifestyles, real efforts at economizing in its use and, perhaps most crucial of all, deciding whether nuclear energy can ever be made safe enough to serve as a long-term substitute for oil, can all determine how long oil will continue to supply our energy needs, and whether the twenty-first century will also be the century of oil – as the twentieth century has been.

GLOSSARY

acid raid   Rain produced when sulphur and nitrogen dioxide from coal fumes and smoke combine with water in the atmosphere.

the Allies   Those countries – principally the USSR, the USA and the countries of the British Commonwealth – who combined together in a military alliance to defeat Germany, Italy and Japan in the Second World War.

anticline   A slightly dome-shaped land formation under which pockets of crude oil are found in regions of the world containing oil-bearing rocks.

apartheid   A social and political system under which the white minority in South Africa rules over the black majority and under which blacks are denied civil rights.

ayatollah   The highest and most spiritually esteemed rank of the clergy in the Shi'a branch of Islam (that branch of Islam which is dominant in Iran).

capitalist   The name given to a system in which all property is privately owned and where production takes place exclusively for profit.

cartel   Any organization of producers who come together to restrict output in order to raise the price of a product.

to caulk   To stop up cracks with a filler.

charterers   In the oil industry, those who hire tankers to carry their oil to its destination.

COMECON   An organization made up of the USSR and a number of east European countries who cooperate closely with each other in planning and linking their economies together.

commodity   Any raw material (oil, cotton, coffee, etc.) which is bought and sold on a regular basis in an international market.

conservative   A political belief that is opposed to fundamental social change.

consortium   A group of producers who come together to share the costs of production and/or to sell their products jointly on a particular market.

coup d'état   The violent overthrow of a government, usually by military means.

cracking   The breaking up of crude oil into its component parts by subjecting it to intense heat and then distilling out the resulting vapours on to cool surfaces within a fractionating column.

deadweight   The difference in the displacement of a ship, loaded and unloaded.

Deep South   Those states in the south-east of the United States where racial discrimination is practised against blacks.

derrick   The upper part or superstructure of an oil well.

embargo   The stoppage of trade; a ban on supplies.

emissions   Outpourings, particularly of smoke or chemical fumes, into the atmosphere, or the spewing of industrial sewage into rivers, lakes and the sea.

environmentalists   People who believe that the natural environment (natural vegetation, clean air and water, etc.) is being endangered and that supplies of non-renewable natural resources like oil are being threatened by excessive industrialization and the over-population of the planet.

exponential rate of growth   A rate of growth in output or demand (or any other measurable quantity) in which each succeeding quantity is a given (often a constant) multiple of a previous amount. An example would be where demand goes up at each stage by doubling or quadrupling itself.

feedstock   A chemical source of manufactured products. Crude oil is a feedstock for petrochemical products.

felaheen   The poor peasants of the Arab countries of North Africa and the Middle East.

fossilization   The process by which animal or vegetable matter is turned into rock by being subjected to immense pressure from material lying above it; a process which takes place over millions of years.

fractions   The various components of crude oil which are separated out in an oil refinery by the process known as *cracking* (see above).

hydro-electric power The generation of electricity by using natural or artificial waterfalls to turn the rotor blades of electrical generators.

imperialist The name given to a state which rules over other states or controls them indirectly. It is also the name given to a believer in imperialistic ideas.

League of Nations An international peace-keeping organization set up after the First World War.

merger The combination of two separate organizations (especially firms) into a single unit.

mullahs Teachers of Islamic religion, its moral code and religious law.

nuclear reactor That part of a nuclear power station in which energy is produced from the transformation of plutonium or uranium atoms under controlled conditions.

OPEC The Organization of Petroleum Exporting Countries; a group of oil-exporting countries, founded in 1960, who got together in 1973 to fix both the price and the output of oil on world markets.

Palestine Liberation Organization An organization which believes that Palestine (or Israel) should become a state ruled by the Arabs who live there.

passover The Jewish festival commemorating the deliverance of the Israelites from the Egyptian angel of death (according to a Biblical story).

pesticides Chemicals designed to kill or deter pests and insects

petrochemicals The wide range of chemicals (e.g. detergents) produced from crude oil.

petroleum That component of crude oil mostly used as motor fuel.

pharmaceuticals Those chemicals derived from an oil base which are particularly used in medicine.

photochemical smog Changes in the chemical composition of the atmosphere caused by the interaction of light rays with industrial fumes and factory chimney smoke.

pitch A black shiny material, the residue of distilled tar.

posted price The common selling price for oil agreed by the major oil companies under the Achnacarry Agreement of 1928.

radical A political belief that favours fundamental social change.

radioactivity The emission of radio waves which occurs when certain atoms (e.g. radium, thorium, uranium) spontaneously break up.

river delta A river estuary in which deposits of natural material brought downstream by the river are deposited, forming obstacles to the rapid flow of the river.

sanctions Measures taken by one country against another to change the policies of that country, whether internally or in the field of foreign affairs.

SASOL process A technical process for distilling oil from coal; named after the town of Sasolberg in South Africa where the process was first developed.

sedimentary rock Rocks laid down in layers and formed from the fossilized remains of tiny sea creatures.

sheik In Muslim countries, the head of an Arab tribe or village; a venerable old man; a religious leader.

solar energy Energy produced by using the sun's rays.

spoil-heaps The often rather ugly heaps of earth containing particles of coal or coaldust (or any other mineral) deposited near mine shafts in mining districts.

uranium A chemical element particularly important in the generation of nuclear power.

Western materialism The excessive concern for material goods (especially at the cost of moral and spiritual values) said by their critics to be displayed by Western industrial countries.

Zionism The belief that the Jewish people should have a state of their own in Palestine.

BOOKS FOR FURTHER READING

British Petroleum, *Statistical Review of World Energy* (annual)

A. Hamilton (ed.), *Oil: The Price of Power*, Michael Joseph Rainbird, for Channel 4, 1986

P. Odell, *Oil and World Power*, Penguin, seventh edition 1983

A. Sampson, *The Seven Sisters*, Hodder and Stoughton, 1975

C. Tugendhat and A. Hamilton, *Oil: the Biggest Business*, Eyre and Spottiswoode, second edition 1975

L. Turner, *Oil Companies and the International System*, Allen and Unwin, second edition 1983

The education and public relations departments of the major oil companies are useful sources of materials for the classroom. They will provide maps, diagrams, illustrations and booklets on more specialized aspects of the oil industry, as well as general materials.

Contact:

Shell Education Service, Shell UK Limited, Shell-Mex House, Strand, London WC2R 0DX.

B.P. Educational Service, P.O. Box 5, Wetherby, West Yorkshire LS23 7EH.

INDEX